# Sky Dance

## The Story of an Introduced Species

Written by Isabel Thomas

Illustrated by Chiara Fedele

| London Borough of Enfield | |
| --- | --- |
| 91200000822861 | |
| Askews & Holts | 14-Oct-2024 |
| J598.863  JUNIOR NON- | |
| ENPOND | |

# OXFORD
## UNIVERSITY PRESS

Great Clarendon Street, Oxford, OX2 6DP, United Kingdom

Oxford University Press is a department of the University of Oxford. It furthers the University's objective of excellence in research, scholarship, and education by publishing worldwide. Oxford is a registered trade mark of Oxford University Press in the UK and in certain other countries

Text © Oxford University Press 2024

Illustrations © Chiara Fedele 2024

The moral rights of the author have been asserted

First published 2024

All rights reserved. No part of this publication may be reproduced, stored in a retrieval system, or transmitted, in any form or by any means, without the prior permission in writing of Oxford University Press, or as expressly permitted by law, by licence or under terms agreed with the appropriate reprographics rights organization. Enquiries concerning reproduction outside the scope of the above should be sent to the Rights Department, Oxford University Press, at the address above.

You must not circulate this work in any other form and you must impose this same condition on any acquirer.

British Library Cataloguing in Publication Data

Data available

ISBN: 978-1-382-04087-7

10 9 8 7 6 5 4 3 2 1

The manufacturing process conforms to the environmental regulations of the country of origin.

Printed in China by Golden Cup

**Acknowledgments**

Series Editor: James Clements

Written by Isabel Thomas

Illustrated by Chiara Fedele

The publisher wishes to thank Llŷr Davies for his advice during the development of this book.

Author photo courtesy of Elodie Guige

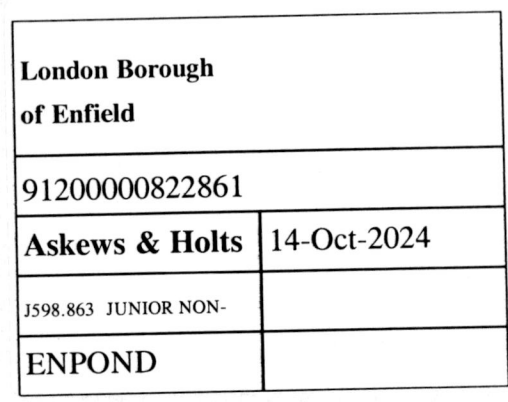

FSC
www.fsc.org
MIX
Paper | Supporting responsible forestry
FSC™ C110497

Starlings are beautiful garden birds, with shimmering feathers and startling calls.

pointed yellow beak

speckled **plumage**

**iridescent** feathers

angular wings

pink feet

short tail

common starling
*Sturnus vulgaris*

Common starlings can be spotted all around the world, from the frozen edges of the Arctic Circle to the **humid** rainforest of eastern Australia.

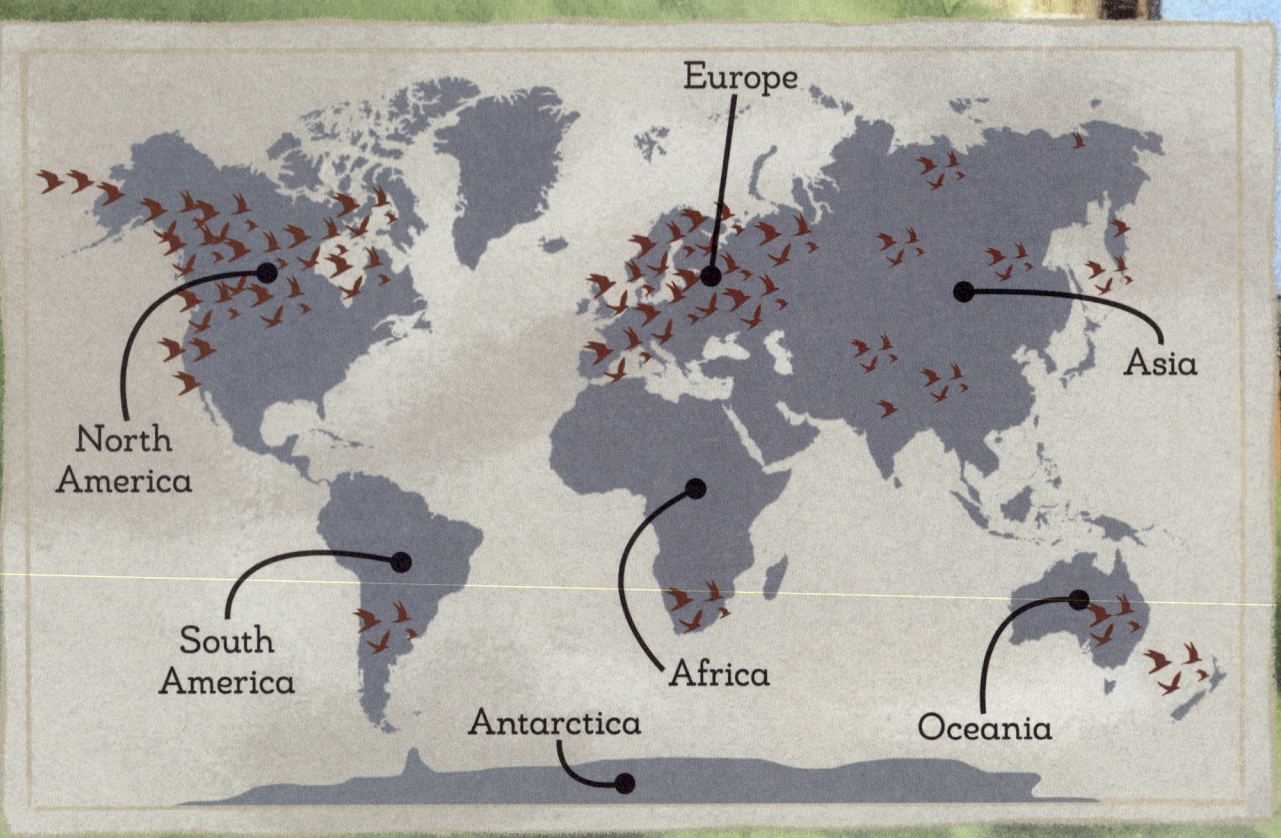

Some of the biggest flocks are found in the United States, which is home to more than 200 million starlings.

However, this was not always the case.

Two hundred years ago, not a single starling lived in North America.

Starlings are too small to fly across vast oceans.

How did they reach new continents?
How did they survive in new habitats?

This is the starlings' story. It begins more than 130 years ago, at the end of the 1800s.

At that time, thousands of people from northern and western Europe were moving to the US every year, from countries such as Great Britain, Norway, Germany, and Ireland.

The European **immigrants** arrived in New York City ready to start new lives. Some of them missed the plants and animals they had grown up with. They set up clubs to ship some of these plants and animals over to North America.

The clubs released thousands of European birds into the skies. Sometimes, they chose birds that helped farmers back in Europe. For example, birds that would eat the caterpillars nibbling on farmers' crops.

Sometimes, they simply chose birds that looked or sounded beautiful.

However, few of the birds survived long enough to lay eggs of their own. They could not find the food or shelter they needed in their new habitats.

In 1890, on a cold winter's day, 80 European starlings were released in Central Park, New York City.

Almost at once they soared into the sky ...

rising and gliding,

diving and dancing,

*darting and parting,
retreating, wings beating.*

This giant space in the middle of New York City seemed like a strange choice. The park was surrounded by buildings and busy roads. There were few trees compared to the countryside, and winters were bitterly cold.

But starlings have some secrets that made them capable of surviving.

Compared to other birds of a similar size, starlings have ...

forward-facing eyes, allowing them to judge distances

big brains for solving problems

small but strong bodies

strong beaks that can push into the ground and **pry** open holes

These features helped the flock of starlings to find insects and other hidden food in the city park, and avoid predators such as cats.

The starlings also found spacious holes to shelter in near the park, such as in the roof of the American Museum of Natural History. Here, they were protected from the cold winter.

A pair of starlings can nest twice a year, laying four to six eggs each time. When the eggs hatch, the chicks are protected from danger in their nests. Their parents bring them insects, millipedes, spiders, and earthworms to eat.

Three weeks after hatching, the baby birds are big enough to leave the nest. They join a flock and will one day have chicks of their own.

Year by year, as the city grew,
the flocks grew too ...

swooping and swirling,

flying and twisting,

swishing and whirling,

whisking the air.

Collecting, connecting, soaring and tumbling,

so close without crashing, painting patterns across the skies.

Thousands of starlings flying together in a huge flock is called a murmuration. Starlings take part in these **aerial** acrobatics by reacting to the movements of the six or seven closest birds. But nobody can say for definite why they do it. Perhaps it is a way to warm up before **roosting**. It could be to confuse or intimidate predators, by being part of a huge crowd.

Wherever flocks found themselves – from city streets to farmers' fields – they were able to quickly find food.

Soon the starlings had spread far beyond New York City.

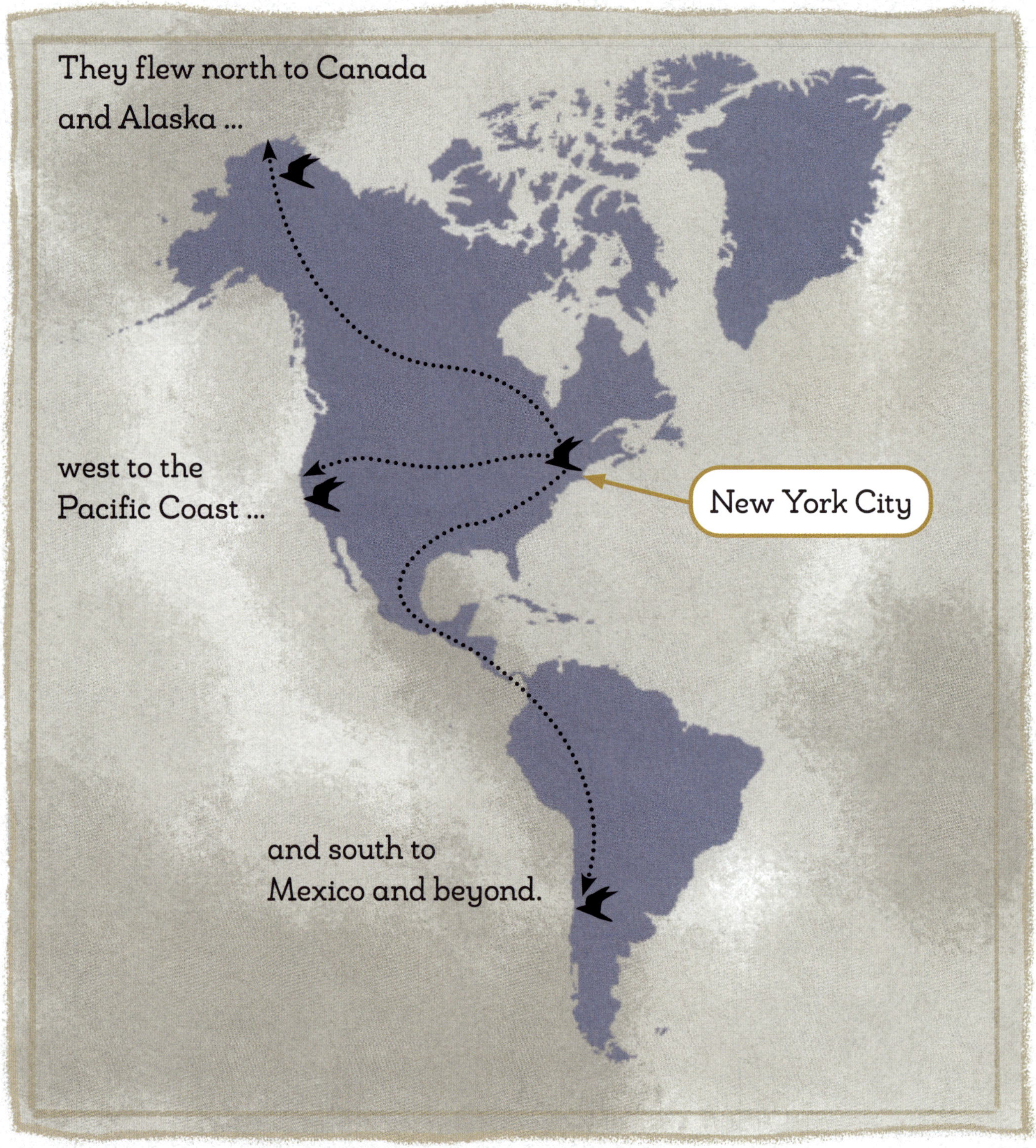

They flew north to Canada
and Alaska ...

west to the
Pacific Coast ...

New York City

and south to
Mexico and beyond.

As starlings **thrived** and filled the skies, some people
changed their minds about them.

They began to see these bold birds as pests.

Returning to roost,
millions of starlings gathered
and chattered and shattered the peace.

They splattered the streets with
germ-laden rain.

They plucked fruit from trees. They swooped over fields and fed on the grain.

People worried that starlings were stealing nest spaces needed by red-bellied woodpeckers and other native birds.

In some parts of the US, people tried to intervene
to control the number of starlings.

They tried ...

tying toy bears to trees

playing owl sounds

flying kites that looked
like hawks

setting off fireworks.

But starlings are brainy birds and they quickly learned not to be afraid. Today, over 200 million European starlings survive and thrive all over the US. All of them are thought to be related to the 80 birds released in Central Park in 1890.

Many people celebrate the beauty of starlings and their murmurations. They argue that starlings have an unfair reputation.

Starlings make a mess in cities, but they have learned to live alongside humans in places where few other **species** can survive.

Starlings steal nesting holes from woodpeckers and other birds, but scientists have found that this has not affected the numbers of all native birds. Starlings are also happy to use spaces in buildings.

If it's available, starlings will eat grain
that is grown to feed humans and farm
animals. But they mostly eat insects,
including some that are pests to crops.

Starlings are thriving in North America.
In Europe, starling numbers are falling.
In the UK, there are less than half
as many starlings as there were
50 years ago. There are many
possible reasons, including
hotter, drier summers
which result in fewer
**invertebrates** for
starlings to
eat, and fewer
suitable places
for starlings
to nest in.

Things that people do to disrupt nature can have unexpected consequences, but we can learn from the starlings' story.

By restoring forests, protecting rivers, and rewilding the spaces we share, we can help all plants and animals to survive and thrive.

Starlings are not the only species that humans have introduced to new habitats. Let's examine some other living things that are classified as **invasive species** in certain areas of the world.

## mammals

raccoon

domestic cat

European rabbit

## birds

sacred ibis

common myna

Canada goose

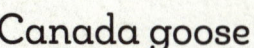

[Not to scale]

## amphibians and reptiles

cane toad

common coquí

Burmese python

## minibeasts

giant African land snail

Asian long
horned beetle

brown marmorated
stink bug

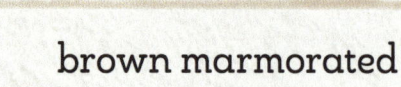

## sea creatures

zebra mussel

European green crab

American comb jelly

Over the past 500 years, the plants and animals that people
have introduced to new areas have caused hundreds of other
species to become **extinct**.

# Glossary

**aerial**: something that is happening in the air

**extinct**: when there are no living members of a species left

**humid**: when there is lots of water vapour in the air

**immigrants**: people who go to live permanently in a country that is not the country they were born in

**invasive species**: living things that humans have introduced to a place where they are not naturally found, and which have an impact on other species in that place

**invertebrates**: animals that don't have a spine or backbone, such as butterflies, spiders and worms

**iridescent**: bright colours that seem to change in the light

**plumage**: all the feathers on a bird's body

**pry**: to open, lift or move something by putting one end of a tool (such as a beak) under it, and pushing down or pulling up on the other end

**roosting**: when flying animals settle down to rest or sleep

**species**: certain types of living things; members of the same species can breed with each other

**thrived**: did very well